Amtrak Train Derailment
Nodaway, Iowa

Investigated by: John Lee Cook, Jr.

This is Report 143 of the Major Fires Investigation Project conducted by Varley-Campbell and Associates, Inc./TriData Corporation under contract EME-97-CO-0506 to the United States Fire Administration, Federal Emergency Management Agency.

Homeland Security

Department of Homeland Security
United States Fire Administration
National Fire Data Center

U.S. Fire Administration Fire Investigations Program

The U.S. Fire Administration develops reports on selected major fires throughout the country. The fires usually involve multiple deaths or a large loss of property. But the primary criterion for deciding to do a report is whether it will result in significant "lessons learned." In some cases these lessons bring to light new knowledge about fire--the effect of building construction or contents, human behavior in fire, etc. In other cases, the lessons are not new but are serious enough to highlight once again, with yet another fire tragedy report. In some cases, special reports are developed to discuss events, drills, or new technologies which are of interest to the fire service.

The reports are sent to fire magazines and are distributed at National and Regional fire meetings. The International Association of Fire Chiefs assists the USFA in disseminating the findings throughout the fire service. On a continuing basis the reports are available on request from the USFA; announcements of their availability are published widely in fire journals and newsletters.

This body of work provides detailed information on the nature of the fire problem for policymakers who must decide on allocations of resources between fire and other pressing problems, and within the fire service to improve codes and code enforcement, training, public fire education, building technology, and other related areas.

The Fire Administration, which has no regulatory authority, sends an experienced fire investigator into a community after a major incident only after having conferred with the local fire authorities to insure that the assistance and presence of the USFA would be supportive and would in no way interfere with any review of the incident they are themselves conducting. The intent is not to arrive during the event or even immediately after, but rather after the dust settles, so that a complete and objective review of all the important aspects of the incident can be made. Local authorities review the USFA's report while it is in draft. The USFA investigator or team is available to local authorities should they wish to request technical assistance for their own investigation.

This report and its recommendations were developed by USFA staff and by Varley-Campbell & Associates, Inc. Miami and Chicago, its staff and consultants, who are under contract to assist the Fire Administration in carrying out the Fire Reports Program.

The Federal Emergency Management Agency, U.S. Fire Administration gratefully acknowledges the cooperation of the Adams County Sheriff's Office, Emergency Management Agency, and Emergency Medical Service as well as the Fire Chief and members of the Corning, Iowa Volunteer Fire Department. Everyone who assisted in the preparation of this report was generous with his or her time, expertise, and counsel.

For additional copies of this report write to the U.S. Fire Administration, 16825 South Seton Avenue, Emmitsburg, Maryland 21727. The report is available on the Administration's Web site at http://www.usfa.dhs.gov/

U.S. Fire Administration

Mission Statement

As an entity of the Department of Homeland Security, the mission of the USFA is to reduce life and economic losses due to fire and related emergencies, through leadership, advocacy, coordination, and support. We serve the Nation independently, in coordination with other Federal agencies, and in partnership with fire protection and emergency service communities. With a commitment to excellence, we provide public education, training, technology, and data initiatives.

TABLE OF CONTENTS

AMTRAK TRAM DERAILMENT
NODAWAY, IOWA
MARCH 17, 2001

Local Contacts: Donnie Willett, Jr., Fire Chief
Corning Iowa Volunteer Fire Department
Corning, Iowa 50841
641-322-4112

Merlin R. Dixon, Sheriff
Adams County, Iowa 50841
Corning, Iowa 50841

Brian Kannas, Emergency Management Coordinator
Adams County Iowa
PO Box 407
Corning, Iowa 50841
641-322-3798

David Walter, Director
Adams County Emergency Medical Service
Corning, Iowa 50841
641-322-3121

OVERVIEW

On the night of March 17, 2001 Amtrak's westbound California Zephyr passenger train derailed at 11:40 PM between Brooks and Nodaway in Adams County, Iowa. The train was traveling along a section of track owned by the Burlington Northern Santa Fe Railway and was carrying 225 passengers and sixteen crewmembers. The maximum passenger capacity is 519. One passenger died as a result of blunt force trauma sustained during the derailment and ninety-six others required transportation to a medical facility. Two of the injuries were serious enough to warrant evacuation by helicopter.

The derailment occurred in a very remote area just east of the Nodaway River. Two engines and nine of the fifteen cars left the tracks, but did not catch fire or spill any fuel. Two of the cars came to rest perpendicular to the tracks and several of the rail cars overturned or came to rest on their sides. Many of the passengers were already asleep when the incident occurred and were disoriented when they awoke because of the attitude of the cars and almost total darkness due to the absence of electrical power.

The crew of the Zephyr contacted the BNSF dispatcher in Fort Worth, Texas and reported that the train had derailed. The SNSF dispatcher contacted the Adams County Sheriff's Department and provided the authorities with the location of the derailment, but was unable to provide any additional details concerning the number and extent of the injuries or the magnitude of the incident. Local fire, EMS, and law enforcement agencies were immediately dispatched.

When the derailment occurred, the Corning Volunteer Fire Department was hosting its annual fund-raiser at the National Guard Armory and was preparing to serve breakfast at the fire station after the festivities concluded. Consequently, the department was fully staffed and the response to the incident was immediate. Upon hearing the incident dispatched, other area emergency services offered their assistance, which was readily accepted. The spouses of the firefighters also went into action to assist with the potential influx of patients at the local hospital and to shelter and feed the passengers that had not been injured.

The scene that greeted rescuers upon their arrival was eerily similar to the movie Field of Dreams. As soon as the passengers spotted the lights from the emergency vehicles, they began to quietly move toward the light and seemed to magically appear out of the darkness. Rescuers commented that the victims were very calm and the incident scene was unusually quiet. Rather than panic, people had attended to the injured and had helped each other evacuate the wreckage.

Access to the derailment site was limited since the derailment had not occurred at a crossing. Open pasture and farmland bordered the site on the north and south. Originally, there had been parallel tracks along the section where the derailment occurred. The southern track had been abandoned, but the roadbed had been maintained by the railroad to provide access along the track and small vehicles were able to use the roadbed to reach the site of the incident from a narrow road that was approximately one quarter of a mile west of the incident site and from another road that was located three-eights of a mile east of the site. Therefore, emergency responders were able to access the derailment site from two directions.

Access from two points proved to be significant, because two of the derailed cars completely blocked both the track and the abandoned roadbed, essentially dividing the incident site in half. Terrain and soft ground prevented the rescuers from driving around the wreckage and the roadbed was so narrow that it was impossible to turn around. Therefore, emergency vehicles were forced to drive in to the site from both the east and west side of the incident and then back out to the road.

Because of the large number of passengers and crewmembers (241), many of those who had not been injured or had only sustained minor injuries were taken out in the back of a pick-up truck or in a four-wheeled drive vehicle. Ambulances were then free to transport the more seriously injured victims to area hospitals following triage and stabilization. The two most seriously injured were taken by helicopter to a trauma center.

Those that had not been injured were taken by school bus to the Nodaway Community Center, where a temporary shelter had been opened. The occupants of the train were re-examined at the Community Center to make sure that they had not been injured. Later into the incident, Amtrak arranged for everyone to be transferred to hotels in Omaha, Nebraska, some seventy miles away.

Approximately 200 emergency responders from a number of local and State agencies responded to the incident. The remoteness of the location allowed law enforcement officials to quickly establish a perimeter to secure the area and to prevent the curious from interfering with rescue efforts. The time of the day also contributed to limiting the number of curiosity seekers at the scene. Most of the civilians that did go to the scene used their personal vehicles (pickups, SUV's, and van's) to assist in the removal of the passengers from the derailment site.

The incident lasted approximately three hours and by the time the national and international media began to descend on Corning, the event was largely over. More than 519 calls from the media and family members were processed by the Adams County Sheriff's Office.

More than fifty investigators from State and Federal agencies scoured the site of the derailment to determine the cause of the incident. The exact cause had not been determined at the time that this report was written, but the National Transportation Safety Board determined that the incident occurred at the site of a broken rail. A section of the rail had been previously removed and had been patched, a common practice in the railroad industry. It was not immediately determined in the derailment occurred because of the broken rail or if the rail broke as a result of the derailment.

SUMMARY OF KEY ISSUES

Issues	Comments
Access	Access was extremely limited due to the remoteness of the accident site, which made the evacuation of passengers and the injured very difficult. On the other hand, the remoteness of the site assisted law enforcement with crowd and traffic control efforts.
Communications	The volume of radio traffic quickly overwhelmed the two-way communications system. The absence of a repeater system and the remote location hampered the use of portable radios and was not conducive to the use of cellular telephones. There were only five incoming 9-1-1 lines and 9-1-1 telephone lines, which were quickly overcome by the 519 telephone calls processed by the Adams County Communication Center.
Media	The remote location and inaccessibility of the incident site allowed the emergency responders to seal off the perimeter and prepare for the arrival of the media. Nevertheless, media inquires were abundant, some as far away as London and Brisbane.
Pre-Incident Planning	While an incident of this type had not been pre-planned, the presence of an up-to-date County Emergency Operations Plan, which had been regularly exercised, contributed to the successful outcome of the incident. The Plan included an annex for tornado response, which can include multiple causalities. Emergency Responders successfully followed the guidelines of the tornado plan in conducting search and rescue operations as well as treating, transporting, and sheltering the large number of injured passengers. The local firefighters had not been trained to rescue and extricate people from trains. Therefore, they relied upon their training in searching for and rescuing victims of tornadoes.
Rescue Efforts	Many of the doors jammed when the derailment occurred and the train lost electrical power. The windows were not breakable and a number of the cars were two stories in height with narrow stairways. All of these factors delayed and hampered in the search and rescue effort. The train derailment occurred at 11:40 hours and many of the passengers on the train were already asleep. The incident site was in an unlit area and the absence of emergency lighting on the train made and compounded the disorientation of the passengers as they attempted to find their shoes and coats and exit the train. Emergency responders were required to illuminate the scene to the best of their ability to conduct search and rescue operations.
Time of Day	The temperature at the time of the incident was 28°F. Winds were calm and the sky was clear. It had recently snowed, but the accumulations on the roadways had already melted. The lack of a wind chill no doubt prevented any cold related temperatures, while the freezing temperature helped stabilize the unpaved roadbed. The cold weather also made it imperative to open a shelter to accommodate the large number of passengers that were not injured, but were exposed to the freezing temperatures.
Weather	If the weather had been warmer the melting snow and ice would have led to the formation of mud on the unpaved access to the incident site, which would have impeded rescue efforts. Elevated conditions would have also contributed to the possibility of heat related injuries to passengers and emergency responders.

THE CORNING VOLUNTEER FIRE DEPARTMENT

Corning is located east of Omaha, Nebraska in the southwestern corner of Iowa and is seventy-one miles west, southwest of Des Moines, the State capital. The city is the county seat of Adams County and has a population of 2,000. Adams County encompasses 432 square miles and the principle occupation of the county's 4,800 residents is agriculture.

The Corning Volunteer Fire Department is one of three volunteer fire departments in Adams County. There are no paid departments within the County. Corning VFD's thirty members provide fire protection within the city limits of Corning and respond to a rural portion of the county as well. The Department operates one fire station located in downtown Corning with four pieces of apparatus. Operations personnel are divided into three teams of firefighters. Each team is lead by a Captain and a Lieutenant who is responsible for training his or her team. Average response time is three minutes within the city limits and the Department averages fifteen members per incident, which includes daytime calls. The Department's response goal is to be able to respond with at least two units and eight firefighters on every call.

The Adams County EMS Service, staffed by both volunteer and career members, provides EMS transport service within the city and the county. The service is based at the Alegent Health Mercy Hospital in Corning. At least two BLS ambulances are staffed around the clock whenever it is possible to do so.

AMTRAK

The name Amtrak is the blending of the words "American" and "Track". The railroad's official name is the National Railroad Passenger Corporation. Amtrak officially began service on May 1, 1971 with a schedule of 184 trains, serving 314 destinations. Amtrak took over the passenger operations of all but three railroads that continued their own intercity passenger train service for a period of time. They were the Rock Island Railroad, the Southern Railway, and the Denver & Rio Grande Western Railroad. The Southern Railway ceased operations of its Southern Crescent in 1979. Amtrak assumed the route, renaming it the Crescent. The Denver & Rio Grande Western Railroad ceased passenger train operations in 1983. Amtrak re-routed its California Zephyr to cover the Denver & Rio Grande's routes.

Since the beginning, even-numbered trains have traveled north and east. Odd numbered trains travel south and west. Among the exceptions are Amtrak's Surfliners, which use the opposite numbering system inherited from their former operator, the Santa Fe Railroad and some Empire Corridor Trains. Amtrak serves more than 500 stations in forty-five States. Those not included are Alaska, Hawaii, Maine, South Dakota and Wyoming. Wyoming is served by Amtrak Thruway Motorcoaches and plans are being developed to serve Maine.

Amtrak operates over more than 22,000 route miles with 25,000 employees. It owns 730 route miles, about 3% of the total nationwide, primarily between Boston and Washington, DC, and in Michigan; eighteen tunnels, and 1,165 bridges. In other parts of the country, Amtrak trains use tracks owned by freight railroads. On weekdays, Amtrak operates up to 265 trains per day, excluding commuter trains. These trains include 100 trains in the Amtrak Intercity business unit and 48 trains in the Amtrak West business unit (Washington, Oregon, and California). Amtrak trains operate every minute of the entire year.

Amtrak operates 2,188 railroad cars including 173 sleeper cars, 743 coach cars, 66 first class/business class cars, 65 dormitory/crew cars, 65 lounge/cafe/dinette cars, and 83 dining cars. Baggage and Mail and Express cars make up the remainder of the fleet. Amtrak also operates 343 locomotives, 278 diesels and 65 electric.

THE DERAILMENT

At 11:40 PM on the night of Saturday March 17, 2001, the dispatcher on duty at the Adams County Sheriff's Office in Corning, Iowa received a telephone call from the Fort Worth, Texas based dispatcher of the Burlington Northern Santa Fe Railway (BNSF) advising her that Amtrak's California Zephyr passenger train traveling westbound on the BNSF's mainline had derailed between Brooks and Nodaway, Iowa. The BNSF dispatcher had been contacted by the train's engineer via their radio system, but was unable to provide any additional details concerning the incident.

The Sheriff's Office serves as the 9-1-1 answering point for all of Adams County and dispatches for all of the public safety agencies within the County. The dispatcher immediately notified the volunteer fire departments, the County EMS Service, and all available deputies. Given the potential magnitude of the incident, the dispatchers also began to recall the off-duty deputies.

Only one dispatcher was on duty at the time the call was received, but it was time for a shift change and the relief dispatcher was already present. The off-going dispatcher remained in the dispatch center through the duration of the incident to help, and as soon as the incident was dispatched offers of assistance from other agencies began to be received, and all were accepted. Upon hearing the call being dispatched, an off-duty senior reserve officer and the Corning Police Chief responded to the Communications Center and assisted throughout the incident. The five in-coming 9-1-1 lines and the five in-coming telephone lines were immediately flooded and remained busy throughout the incident. A total of 519 calls were processed by the dispatch center.

March 17th is also Saint Patrick's Day and the Corning VFD was hosting its annual fund raising event, a Firemen's Ball, at the local National Guard Armory. After the Ball, attendees were scheduled to adjourn to the fire hall for breakfast and some of the members had already retired to the fire hall to help prepare for the breakfast when the incident occurred. Since the temperature was below freezing, the Fire Chief had instructed members to start the apparatus, which was parked outside on the apron. The Chief wanted the engines to be warm and the windshields to be free of ice or frost in the unlikely event something were to occur during the breakfast.

The sky was clear on the night of March 17. The temperature was 28°F and the wind was calm, thus no wind-chill. These conditions helped prevent hypothermia and prevented emergency responders from becoming overheated during the incident. Had it been colder, there would have been a problem due to the large number of victims and the potential for hypothermia. Victims were awaken from their sleep and were unable to find their shoes or coats due to the darkness and position of many of the rail cars. Colder temperatures, however, might have frozen the ground, which would have made access easier. Nevertheless, freezing temperatures did help keep road passable and prevented them becoming muddy due to the presence of accumulated snow and ice, and the fact that the roads were unpaved. If the event had occurred one week earlier, the results might have been different due to the amount of snow that had been on the ground in the area. The incident would have had a snow removal problem.

Amtrak's California Zephyr, also known as the Number Five, had last stopped in Des Moines, Iowa and was traveling along a portion of track owned and operated by the Burlington Northern Santa Fe Railway (BNSF) when it derailed one and one-half miles west of Brooks and east of Nodaway. The Zephyr originated in Chicago and was destined for Emeryville, California, which is near Oakland.

The train consisted of two locomotives and fifteen cars and its next scheduled stop was Omaha, some seventy miles to the west. The train was confirmed to have been traveling at fifty-two miles per hour on a rail section that is rated at eighty miles per hour. The slower speed was due to problems with the train's whistle, which is used at crossings to warn motorists.

Rescuers were unable to immediately determine the number of people on the train because the conductor had just begun the task of collecting tickets when the derailment occurred, which particularly made it difficult to account for the number of children on board. While the train had the potential to transport 519, the initial reports placed the occupant load at 210 - 195 passengers and fifteen crewmembers. That number was later revised upward to 241 - 225 passengers and sixteen crewmembers.

The first law enforcement officer on the scene was a State Trooper who approached the incident site from the west side. The officer placed the crew of the derailed engine in his car. Railway officials later took the crew to be tested for alcohol and drugs. Law enforcement agencies from throughout the area responded and were primarily used for traffic control. The time of day and the remoteness of location resulted in a very limited problem with spectators. The response by rescuers in their personal vehicles and that of nearby residents contributed significantly to the congestion at the site.

When the derailment occurred, a number of the cars came to rest perpendicular to the track (see Photo One). Two of the cars completely blocked the scene and essentially divided the incident into two sectors, east and west. Wreckage and debris was scattered over an area approximately 1/4 mile in length. Access was limited to a narrow road located 3/8th of a mile east of the wreckage and another road 1/4 mile to the west. Staging and triage areas were established on both sides of the site.

The derailment did not cause a fire and there were no hazardous materials involved. Both engines and a majority of the cars came off the track, however, with at least one being turned upside down and approximately four cars being turned on their side. The train traveled through fences as it left the track, but none of the affected fields contained livestock. Additionally, no utilities were disrupted by the derailment. There, was an old telegraph line along the track, which had been abandoned and it was quickly determined not to be a hazard.

The rail bed at the site of the incident had space for two parallel tracks, although only one set being used at time of the crash and generally ran from northeast to southwest. The section is part of the BNSF's mainline and carries approximately forty freight trains a (day and some 103 million tons of freight annually. The primary product is coal being hauled east from Wyoming. The abandoned roadbed proved to be the only means to each site. Ambulances and rescuers had to drive into the site, in single file and then back out with their patient before another vehicle could enter the site. This process was complicated by the presence of several trestles, which spanned small streams. The trestles were open between the cross ties, thus making crossing on foot very dangerous.

Ninety-six people were injured and one person was killed in the derailment. No one was ejected as a result of the derailment, but access to the victims was limited by the position of the cars and the very narrow stairways on the double-decked cars. Two of the cars came to rest perpendicular to the track and essentially sliced the scene into two sectors. Passenger rail cars are very sturdy and have unbreakable windows. The doors began jammed as a result of the impact of the derailment.

As the emergency responders began to arrive, a scene eerily reminiscent of the movie *The Field of Dreams* greeted them. People began to walk toward them and their lights. The responders also reported an unusual sense of calm, which pervaded the incident scene. Many of the passengers directed rescuers to those with more severe injuries and often declined immediate transportation in order for family members to remain together.

The ninety-six people who were injured were transported to six area hospitals, primarily by ground ambulance. Two of the victims' injuries were serious enough to warrant evacuation by helicopter. A third person was later flown out from the Corning Hospital. The first ambulance initially responded from the Adams County EMS agency and assumed command of the triage efforts. The supervisor and the County's second EMS unit responded to the Corning hospital's emergency room to assist with the influx of patients because of the small capacity of the hospital and the expected patient load. The supervisor helped coordinate the transportation effort and directed ambulances to area hospitals based upon the nature and severity of the injuries.

Due to the large number of injuries, ground ambulances were required to make multiple trips and they began to run out of backboards and c-collars. In many cases the used duet tape to secure patients to stretchers and to immobilize patients. Rescuers used ambulances as well as privately owned vehicles, many of them four-wheel drive pickup trucks and SUV's, to transport the injured and non-injured alike to the staging areas, one each side of the derailment site. At staging, the occupants of the train were triaged and were sorted for transport to a medical facility. If they were not injured, passengers were taken by school bus to the Community Center in Nodaway, which had been opened to shelter the non-injured.

To assist with the rescue effort, a load of linens and blankets were taken to the scene by a vendor of the Mercy Hospital in Corning because of the temperature and the large number of occupants on the train. The number of blankets onboard the ambulances and train, however, proved to be sufficient. The hospital had been notified of the incident fifteen minutes in advance of the receipt of the first patient, which allowed sufficient time to mobilize its staff and kitchen. All four of the local physicians and the entire off-duty nursing staff responded to the emergency room to assist with the influx of patients. A total of 43 patients were taken to Mercy Hospital. Three patients subsequently transferred to metro hospitals and one remained hospitalized in Corning. The remainder were all treated and released.

At Iowa Methodist Medical Center in Des Moines, a 77-year-old female was hospitalized with chest injuries and a fractured wrist. At St. Joseph in Omaha: 54-year-old male and 50 year old female were hospitalized and a 47-year-old female was hospitalized with broken hip at the University of Nebraska Medical Center in Omaha.

As patients arrived and were treated, their names were faxed to the Dispatcher in order to keep track of everyone. The nature and severity of their injury and the location of their transport was noted by the Dispatcher as well.

The single fatality was a 69 year-old female from Colorado Springs that had traveled to Des Moines because her brother, a nursing home resident in Des Moines, had died on the previous Thursday. She was in the process of transporting his ashes to Colorado when the incident occurred. An autopsy was conducted by the State medical examiner and it was determined that she died at the scene from multiple blunt force trauma injuries. The medical examiner ruled that her death was accidental.

Approximately 200 emergency responders assisted at the scene of the incident. Twenty-one ground transportation ambulances, two helicopters, and an undetermined number of private vehicles were used to transport patients. For a complete list of the responders, see Appendix C. The Adams County

Emergency Management Coordinator, which is a voluntary position, served as the incident commander. A Command Post was set up on the Westside of the incident and the Corning Fire Chief directed operations from that site. The Taylor County Emergency Management Coordinator reported to the staging area on the west side of the incident and assisted as needed.

Passengers and crewmembers that had not been injured were transported by school bus to the Nodaway Community Center where they were sheltered from the cold and fed. An EMS unit was place on standby at the Community Center to ensure no one was later discovered to have been injured or if there had been a medical emergency triggered by the delayed effects of the incident. The Emergency Management Coordinator from Union County went there to assist. As the incident progressed, Amtrak transferred them to hotels in Omaha where they received counseling and assistance from the Red Cross and the Salvation Army.

Approximately fourteen minutes into incident the media began to call the Adams County Sheriff's Dispatcher. Phone calls from as far away as London and Brisbane, Australia were answered. Once the news hit the Associated Press newswire, the call volume greatly increased. One local media person reported that the local media representative got in early and the event was really over by the time the big guys got interested. Most of the large agencies arrived after daylight, but by then it was all over except for the investigation and cleanup. Family members of the occupants of the train also began to call once the news of the derailment was released. A number of television stations responded to the area with their satellite trucks and helicopters. There was not need, however, to restrict air space.

The local emergency responders had not had any training on responding to or managing a train wreck and had very limited knowledge about train wrecks and the tactics and logistics necessary to manage such an event. Therefore, they relied upon the training that they had, which is the response to tornadoes that are common in the Midwest. Rescuers conducted both a primary and secondary search of the train and marked the cars as they went using the orange spray paint that is a part of their tornado response plan to mark buildings that have been searched in the event that a tornado hits and does serious damage.

By 07:00 hours on Saturday Morning, most of the passengers had been bused to Omaha hotels and by 09:00 hours only a few volunteers remained at the Community Center. The rail line was reopened to freight and passenger service at 01:45 hours on Monday March the 20th. The temperature and the nature of the rescue activities made rehab unnecessary according to local rescue officials. No injuries of emergency responders were reported during the incident, nor were there any injuries reported by the occupants of the train as a result of the rescue efforts.

THE INVESTIGATION

Approximately fifty investigators from the National Transportation Safety Board (NTSB), the Federal Railroad Administration, the Iowa Department of Transportation, Amtrak, the Burlington Northern Railway, the Brotherhood of Locomotive Engineers, and the United Transportation Union participated in an investigation of the derailment. Investigators concluded that the train left the track due to a broken rail, but were not immediately able to determine if the rail was broken before the incident occurred or if the rail broke as a result of the derailment.

The broken rail was discovered at a point where a patch had previously been made to replace a section of defective rail. A spokesperson from the NTSB indicated that this was standard procedure to put in a patch where a problem has been detected. A maintenance crew had cut a gap into the defective rail and bolted in another piece of rail approximately four yards in length. NTSB personnel collected pieces of the broken rail and sent them to Washington, DC for a metallurgical analysis to determine if there were any defects, which might have contributed to the derailment.

State and local law enforcement officials did not participate in the investigation because the derailment occurred on private property and did not occur at a railroad crossing. Therefore, law enforcement officials indicated that they did not have any jurisdiction in the matter because no crime had been committed. Their primary role at the incident was traffic and crowd control and to assist fire and EMS personnel.

The BNSF track had been visually inspected on the day of the accident and no irregularities had been detected and investigators determined that an eastbound coal train had passed over the derailment site fifty-seven minutes before the derailment without incident. The Amtrak crew was tested for drugs or alcohol following the incident. The results of those tests had not been released when this report was prepared. Repair crews placed a section of temporary track in the vacant roadbed to reroute trains around the derailment site in order to allow the investigators to conduct their investigation undisturbed.

The NTSB praised the local emergency crews for their response effort, as did officials from Amtrak. The NTSB also had praise for Amtrak for placing emergency light sticks on their trains based upon recommendations from the NTSB. NTSB stated that they believed the availability of a light source helped people survive the derailment.

No formal determination as to the exact cause of the derailment had been released at the time that this report was prepared. In a related note, The California Zephyr was involved in a collision with another train on September 13, 2001 in Wendover, Utah, which is near the Nevada line some 120 miles west of Salt Lake City. Only minor injuries were reported, although parts of the train did catch fire.

LESSONS LEARNED

1. **An effective, functional means of communications is essential.**

   There were no turf issues that are so often associated with this type of event. Most of the rescuers and emergency responders knew each other, proving the value of daily verbal, face-to-face communications. Technology, however, was overwhelmed by an event of this magnitude. The Sheriff Department's simplex VHF radio system (155.010 Mhz) was simply overloaded and plagued by dead areas created by the remoteness of the incident site and topography. Cellular telephone also proved to be ineffective because of dead areas and the lack of towers near the incident site.

   Incident planning should include provisions for alternate means of communications. As proven by this incident, small rural communities are not immune to disasters that can quickly overtax an otherwise functional communications system.

2. **Incident Command and Accountability Systems are important elements in the successful management of a large-scale incident.**

   The management of 241 victims and 200 rescuers is difficult under the best of circumstances. An incident command and accountability system is an essential element in the successful management of large-scale events. To be effective, everyone involved must be familiar with the system and should regularly use the system so that it will be automatic in these types of events.

3. **Access and logistics are critical to the success of a large-scare incident.**

It was necessary to perform three extrications during this incident. There was a great need for ropes, hand lights, scene lighting and short ladders because of very limited access due to unbreakable windows, and the narrow stairways in the two-story cars. Prybars, backboards, and c-collars were also in great demand.

There was very limited access to the site. Rescuers had to drive into from two different directions and then back out. The process was complicated by the darkness of the scene and a number of railroad trestles open to the water below.

Transportation was also issue. When the number of ambulances was exhausted, school buses and privately owned vehicles were pressed into service in order to remove passengers and victims to safety.

Future planning for a similar event should identify the types and sources for the equipment that could potentially be required in a multiple casualty incident. Training for the incident should include driving techniques for maneuvering in tight quarters.

4. **Pre-Incident Planning is an important tool in the successful management of a train derailment or other large-scale event.**

Adams County had an often exercised and up-to-date Emergency Operations Plan at the time of the incident. The County activated its Emergency Operations Center and followed the plan. It worked and the incident proved the value of exercising the plan. The general comments received after the incident, including those from Amtrak, were very positive. In the absence of a contingency specific plan, emergency responders relied upon their tornado response plan, which proved to be effective in managing the incident. Debriefings and a critique were held at the community center following the incident. Officials from the Corning hospital and Amtrak coordinated the event.

APPENDIX A

Maps

Map One: Detailed Route Map for the California Zephyr

Map Two: General Route Map for the California Zephyr

Map One: Map One: Detailed Route Map for the California Zephyr

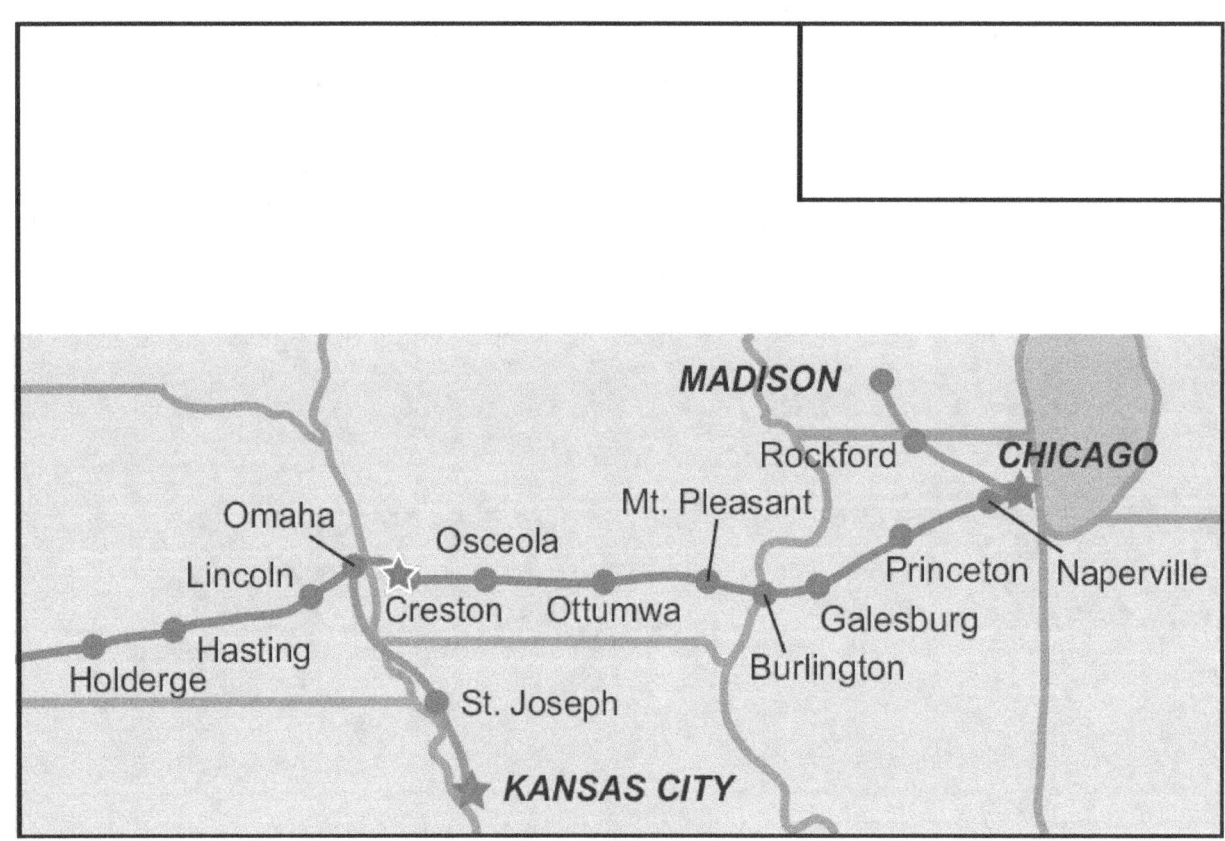

Map Two: Map Two: General Route Map for the California Zephyr

APPENDIX B

List of Photos

Photo	Description
Photo #1:	Aerial view of wreckage
Photo #2:	Photo of the California Zephyr
Photo #3:	Photo taken after the track was reopened showing the position of the wrecked passenger cars and a BNSF freight train passing through the site
Photo #4:	Investigators leaving a wrecked car
Photo #5:	Photo of one of the wrecked cars (a double decker)

1. Aerial view of wreckage

Aerial view of wreckage

2. Photo of the California Zephyr

3.   Photo taken after the track was reopened showing the position of the wrecked passenger cars and a BNSF freight train passing through the site

3.   Photo taken after the track was reopened showing the position of the wrecked passenger cars and a BNSF freight train passing through the site

4.   Investigators leaving a wrecked car

4.   Investigators leaving a wrecked car

5.   Photo of one of the wrecked cars (a double decker)

APPENDIX C

List of Participating Agencies

The following agencies responded to and/or participated in a support role
at the incident on 17 March, 2001

Fire Departments
Corning Volunteer Fire Department
Clarinda Fire and Rescue Department
Creston Fire Department
Lenex Fire and Rescue

Emergency Medical Services
Adams County Emergency Medical Service
Bedford Ambulance Service
Hamburg Rescue Service

Law Enforcement Agencies
Adams County Sheriff's Department
Corning Police Department
Iowa State Police

Other Agencies
Adams County Office of Emergency Management
Alegent Health Mercy Hospital
Corning Community School
Greater Community Hospital
Iowa Department of Public Health/EMS
Iowa Emergency Management Agency
Life-Net
Red Cross
Salvation Army
Taylor County Emergency Management Agency
Union County Emergency Management Agency

APPENDIX D

Recent Passenger Train Derailments

Source: Brotherhood of Locomotive Engineers

- **5 February 2001:** Amtrak train collides with a freight train on the same track in Syracuse, New York injuring sixty-one people

- **4 November 2000:** Amtrak train derails after hitting a truck near Moorpark, California injuring killing the truck driver and injuring thirty people on the train

- **15 March 2000:** Amtrak train derailed alongside a cornfield near Carbondale, Kansas injuring twenty-nine passengers

- **20 September 1999:** Amtrak train rear-ended a freight train in a rail yard in Cumberland, Maryland injuring thirty-seven people

- **16 March 1999:** Amtrak train collided with a truck and derailed near Bourbonnais, Illinois killing thirteen and injuring more than 100

- **9 August 1997:** Amtrak train derailed while crossing a flooded stream in northwestern Arizona injuring 140 people

- **23 November 1996:** Amtrak train jumped the track at sixty miles per hour in Secaucus, New Jersey sideswiping another train and injuring thirty-two

- **16 February 1996:** Amtrak's Capitol limited and a MARC commuter train collide in Silver Spring, Maryland killing all three crew members and eight passengers on the MARC train

- **9 February 1996:** Two New Jersey commuter trains collide killing two crew members and one passenger

- **22 September 1993:** Amtrak's Sunset Limited jumped the rail on weakened bridge that had been rammed by a barge minutes earlier and plunges into a bayou near Mobile, Alabama killing forty-seven

Note: Auto accidents kill 10 times as many as train accidents according to the National Association of Railroad Passengers in Washington, DC. Deaths per 10 billion passenger miles are as follows: 96.2 for cars; 8.6 Amtrak; 3.6 airplane passengers; and 1.4 for buses.

www.ingramcontent.com/pod-product-compliance
Lightning Source LLC
Chambersburg PA
CBHW081249170526
45165CB00009B/3250